End of Term for the Schools Class 1960 - 1962

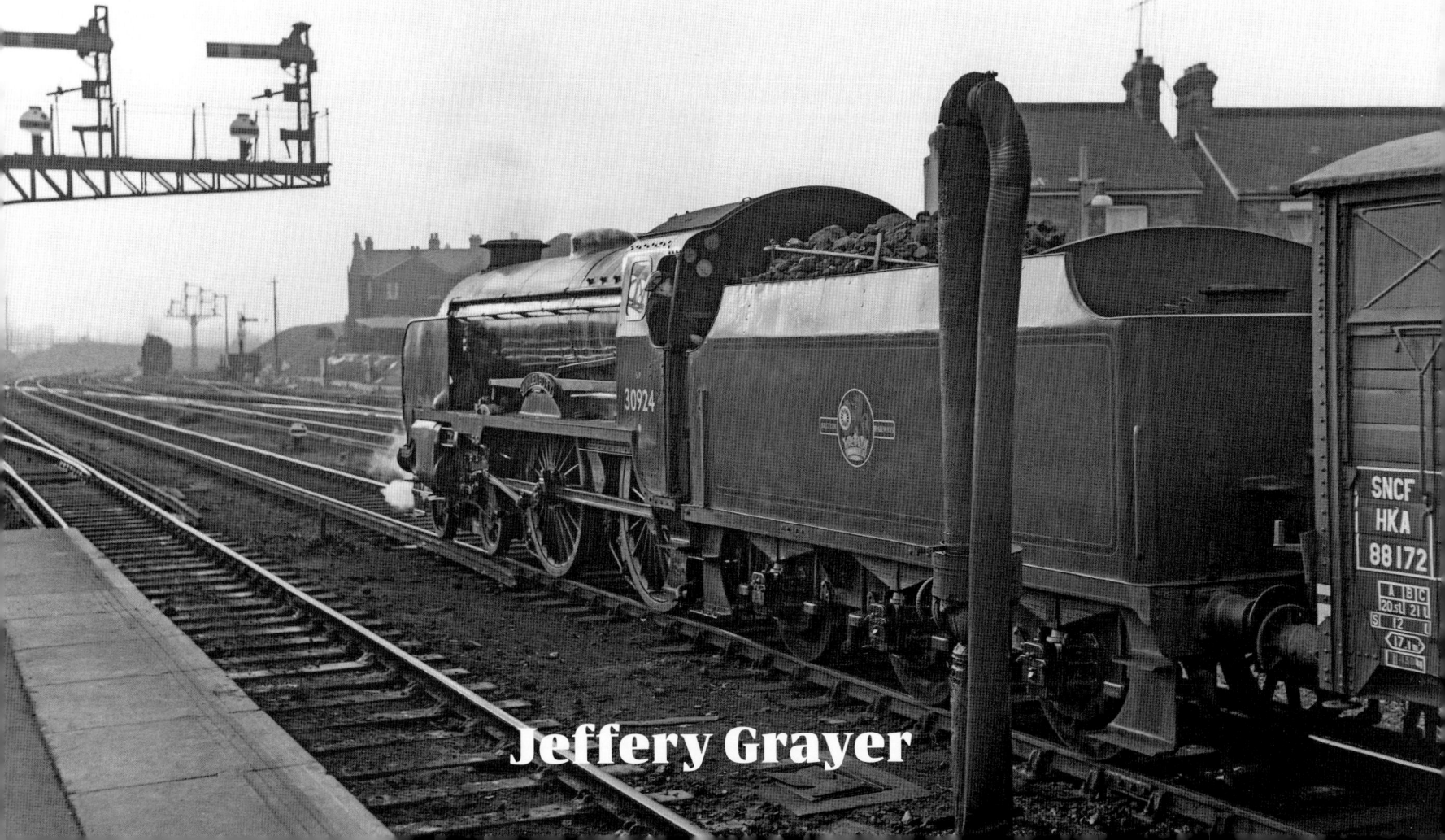

Jeffery Grayer

TOTEM
PUBLISHING

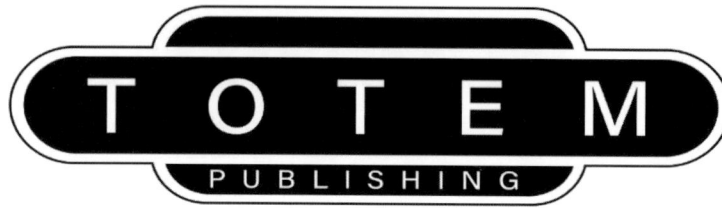

© Images and design: Transport Treasury 2025 Text: Jeffery Grayer

ISBN 978-1-913893-62-0

First published in 2025 by Transport Treasury Publishing Limited. 16 Highworth Close, High Wycombe, HP13 7PJ.
Totem Publishing an imprint of Transport Treasury Publishing.

www.ttpublishing.co.uk

Printed in Tarxien, Malta by Gutenberg Press Ltd.

'*End of Term for the Schools Class 1960 - 1962*' is one of many books on specialist transport subjects published in strictly limited
numbers and produced under the Totem Publishing imprint using material only available at The Transport Treasury.

Front Cover: The 4.40pm service to Brighton via Oxted, Tunbridge Wells and Lewes waits departure time at London Bridge on 3 June 1960. Brighton based Schools Class No 30917 *Ardingly* heads the train which will take no less than 2 hours and 12 minutes to reach its destination via this circuitous route of 62¾ miles. *(Dave Clarke)*

Frontispiece: No 30924 *Haileybury* is held by signals on the up through road at Tonbridge on 2 April 1960 whilst working an up freight service consisting of double deck flat wagons containing imported cars headed by a French Railways SNCF ferry van. *(Dave Clarke)*

Opposite: Spending 4½ years at Brighton (75A) shed at the end of its career, No 30901 *Winchester* was often to be found here in the early 1960s and is seen in steam in this view taken in the shed yard with the impressive trainshed of Brighton terminus in the background. *(Author)*

Rear Cover: The yard of George Cohen saw off the only examples of the class to be completely dealt with by outside contractors. These were Nos 30902 *Wellington*, 30921 *Shrewsbury* and featured here 30935 *Cranleigh* photographed in April 1964 shortly before it and its sisters were reduced to heaps of scrap metal. In this view the 4-4-0 keeps a vigil with S15 No 30507 and LT Underground stock. We should perhaps mention here that the remains of Nos. 30922 and 30931 did find their way to another private scrapyard, that of Wards at Bolton-on-Dearne. This pair had initially been partially cut up at Ashford Works but wagonloads of the resulting debris, which had been stored for a while at Swinton near Rotherham, were eventually consigned to Wards for final disposal. *(Alec Ford)*

Thanks go to Dennis Troughton for his help with the accuracy of this book.

Introduction

Surely one of the most attractive classes of steam locomotive to be found on the Southern Region, as well as what was claimed to be the most powerful 4-4-0 operating in the UK, being the only locomotive of this wheel arrangement to be accorded a 5P power classification. Maunsell's Schools class found a place in the affections of many an enthusiast; a sentiment made more poignant by the rapid withdrawal of the remaining members of the class en bloc in the infamous "locomotive cull" of December 1962.

In this volume we take a look at all forty examples captured during the final three years of their operational existence from 1960 – 1962. For each locomotive, brief details of allocations during this period together with withdrawal, storage and scrapping dates are provided at the head of the page. I was fortunate enough to see many class members during this time as I was a frequent visitor to Brighton depot where several could be observed and witnessing their passage through my home city of Chichester, operating on the through trains from Brighton to Bournemouth and as far as Salisbury on the Cardiff and Plymouth services. Latterly I saw several at Eastleigh depot after their withdrawal and prior to the cutting that went on at the rear of Eastleigh Works. Whilst 1962 marked the end of their working lives, May 1963 witnessed the final movement under its own power of one member of the class, namely No 30934 *St. Lawrence* which ventured down the mainline from Basingstoke to Eastleigh towing a withdrawn Drummond "Black Motor". Even as late as 1964 some could still

be seen, in this case in faraway Northamptonshire, where a trio lingered on in the scrapyard of George Cohen near Kettering.

It is now 95 years since they first appeared on the scene back in 1930, revolutionising travel over the Hastings route, being specifically designed to cope with the infamous narrow bores of its tunnels. Following the Hastings Dieselisation of 1957 and Stage 1 of the Kent Coast Electrification in 1959 many were transferred away from their traditional South Eastern haunts to the Central and South Western Divisions of the SR. Work was found for them on the Reading-Guildford-Redhill cross country service and on inter-regional trains from Kent and Sussex to the Midlands and beyond such as the Eastbourne - Glasgow car sleeper service. They also appeared on former L&SWR main lines to Salisbury and Exeter and to Bournemouth, Weymouth and powering the Lymington boat trains as far as Brockenhurst, but amid much shuffling around between depots and against a context of more modern motive power being readily available, sufficient work could not really be found for them. The class, along with many others, was to be swept away with unseemly haste at the end of 1962 following a diktat that came down from on high to cull much of BR's older steam stock, a move that was generally seen as an accounting ploy implemented prior to the replacement of the British Transport Commission by the British Railways Board at the beginning of 1963. This action was repeated on other BR regions but on the SR it was to see the end of some famous and much loved classes of locomotive in addition to the Schools such as the King Arthurs, Lord Nelsons and Billinton's K class moguls.

In choosing images to illustrate this period I have tried to vary the locations and duties featured and have also included some post 1962 views to illustrate the scrapping process. Fortunately today three examples have been preserved, namely 30925 *Cheltenham* as part of the National Collection currently based on the Mid Hants Railway, 30926 *Repton* based on the North York Moors Railway and 30928 *Stowe* which, after a period on static display at Beaulieu, is now at the Bluebell Railway. In some nine years time this particular trio will be celebrating their centenary, having been constructed in 1934, and will hopefully continue to give pleasure for many years to come.

Jeffery Grayer, Devon 2025.

Having arrived from Basingstoke in steam in May 1963 No. 30934 "St. Lawrence" awaits its fate on the scrap road alongside Eastleigh shed before meeting its "Waterloo" at the adjacent Works in August that year. (Author)

Withdrawn in November 1962 and stored at Stewarts Lane No. 30928 was held back from potential scrapping in May 1963 being subsequently purchased by Montagu Ventures (Lord Montagu) and was transferred from rail to road at Millbrook in February 1964. After arrival at Beaulieu an inauguration ceremony was held on 24th March 1964 at the Motor Museum where it shared the limelight with three Bournemouth Belle Pullman coaches. It subsequently moved to the East Somerset Railway at Cranmore in 1973 before moving to the Bluebell Railway in 1980 where it was returned to working order in 1981.
A full overhaul began in 2013 and is currently ongoing.
It is seen here at Beaulieu in the late 1960s. *(Author)*

30900 Eton

Allocated Brighton from 6/58 until 2/62. Stored Brighton 10/60 - 2/62.
Withdrawn 2/62. Scrapped Ashford Works 3/62.

On Saturday 27 August 1960 the through express service from Walsall to Hastings approaches Preston Park to the north of Brighton with doyen of the Schools class No 30900 *Eton* at its head. In the distance on the up line an EMU disappears northwards whilst another EMU occupies one of the berthing sidings seen on the left. This was the final Saturday on which this short dated service operated that year, having commenced on 25 June. Departure from Walsall was at 11:05 reaching Brighton at 3:27pm after having called at Birmingham New Street and Coventry. Hastings was reached at 5:01pm after calls at Eastbourne, Bexhill and St. Leonards. *(Brian Wadey)*

In September 1960 No 30900 enters Redhill from the south passing "B" signalbox with an inter-regional service from Hastings to Northampton, Market Harborough and Leicester (London Road). The lines to the left led to Tonbridge whilst those to the right served Reigate, Dorking and Guildford. The signalbox was to an SER design and although not visible in this shot was protected by a heavy blast wall during WW2 serving as an ARP base during the conflict. It was abolished in 1985 consequent upon the introduction of remote signalling from Three Bridges. *(Roy Hobbs)*

On 28 May 1960 No 30900 passes through Hove with a westbound service either heading for Bournemouth or one of the inter-regional services to Cardiff or Plymouth as the coaching stock has carriage boards attached. The leading coach set, No 885, is one of the Mk 1 four car corridor sets introduced in 1952 seen here in "blood and custard" livery which in November 1960, would be repainted in green. At the other platform a 2 BIL EMU awaits departure for Brighton. The covered footbridge was primarily a public footpath as a subway was provided for inter platform transfers. The elaborate porte-cochere seen on the right was one of two provided here for both the 1865 station building seen beyond the footbridge and for the later 1893 building seen nearest the camera. *(Leslie Freeman)*

Minus connecting rods and nameplates but still retaining its smokebox number a forlorn looking *Eton* awaits the inevitable on the scrap line at Ashford Works yard on 25 February 1962. Between them Ashford and Eastleigh Works cut up the majority of the class with just three escaping to be dismantled by outside contractors with a further trio making it into preservation. This is not quite the whole story however as following scrapping at Ashford "chunks" of semi-cut Nos. 30922 and 30931 were transported by rail in wagons to a firm in Swinton near Manchester for further cutting. However, the remains were sold on to T W Ward of Bolton-on-Dearne who finally disposed of them. *(Leslie Freeman)*

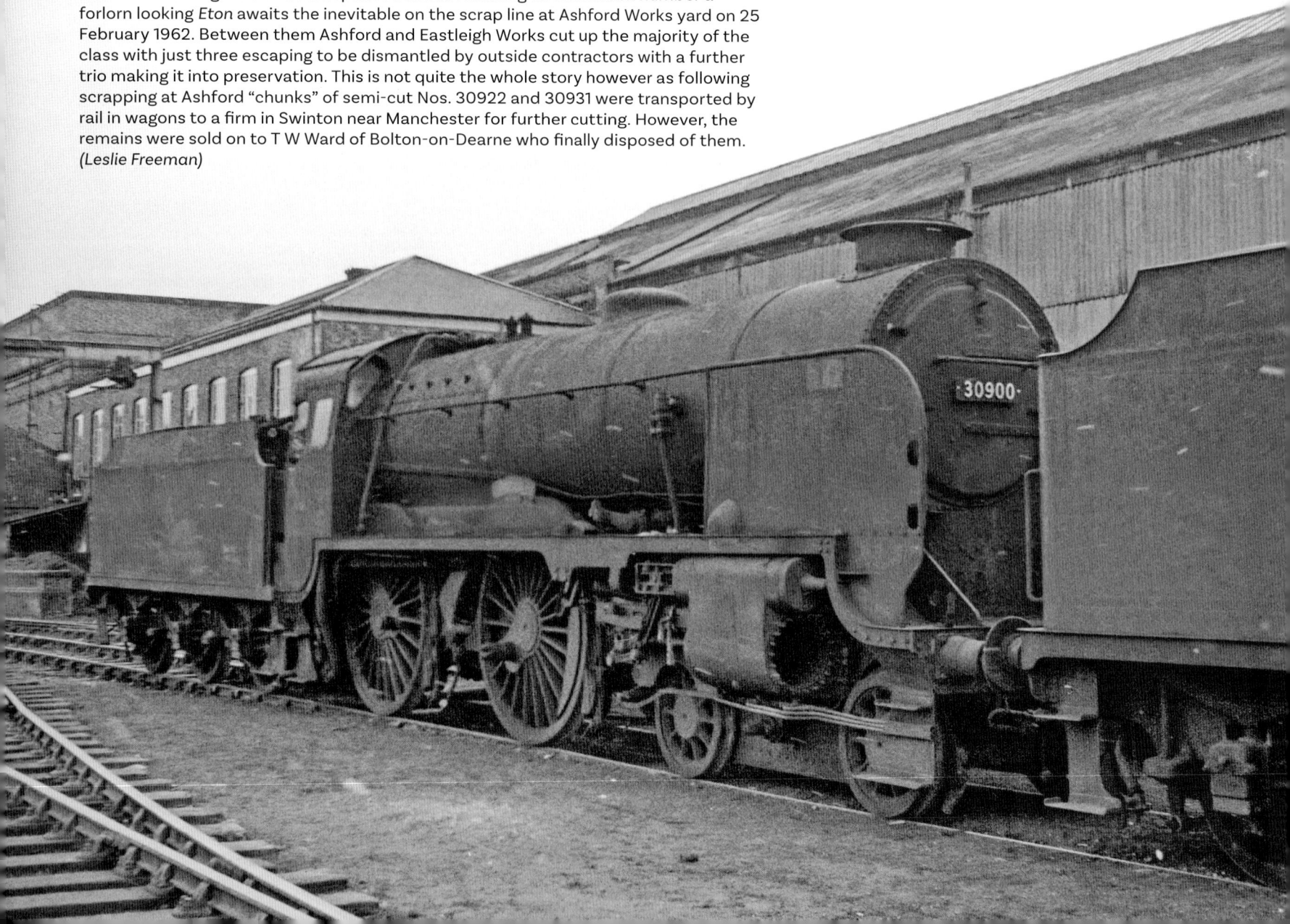

30901 Winchester

Allocated Brighton from 6/58. Withdrawn 12/62. Stored Hove Goods Yard 12/62 – 6/63. Scrapped Eastleigh Works 8/63.

Looking resplendent No 30901 poses at Ashford on 30 June 1960 after the Works had completed a general overhaul on the locomotive nineteen days before. This work had involved the replacement of all tubes, the installation of a speedometer, the part fitting of AWS equipment and repainting into the attractive green livery which seemed to suit this class so well. *(R. C. Riley)*

A scene I well remember, a Schools taking water at my home city of Chichester whilst carrying the headcode denoting the through train from Brighton to Bournemouth. This was due into Chichester at 10:25am and reached Bournemouth West at 12:22pm after calls were made at Southampton Central, Brockenhurst, New Milton, Christchurch, Pokesdown, Boscombe and Bournemouth Central. This useful through train was withdrawn in the winter of 1961/2 thereafter running during the summer only until it was permanently withdrawn from September 1964. *(Roy Vincent)*

Heading south down the main Brighton line and passing Copyhold Junction to the north of Haywards Heath comes No 30901 *Winchester* on Sunday 1 April 1962 with a freight the headcode of which denotes a service emanating either from Bricklayer's Arms or London Bridge and heading via Redhill's quarry avoiding line to either Eastbourne or Hastings. The lines on the right are the branch to Horsted Keynes which was double tracked and somewhat surprisingly electrified in July 1935 with a shuttle service operating to Seaford until withdrawal of passenger services in 1963. *(Brian Wadey)*

30902 Wellington

Allocated Nine Elms 6/58 – 5/60. Brighton 5/60 – 11/60. Nine Elms from 11/60. Withdrawn 12/62. Stored Nine Elms 12/62 – 3/64. Scrapped Cohen's Kettering 4/64.

No 30902 arrives at a rain soaked platform 7 of Lewes station on 3 December 1960 with a service for Brighton which had originated from Tonbridge and travelled south via Eridge. This impressive eight platform station catered for trains from and to Brighton on platforms 6-8. The platform on the far left numbered 4/5 was a single track with two platform faces but was subsequently taken out of use and infilled. *(Leslie Freeman)*

No 30902 has just arrived at Yeovil Town on 30 September 1961 having brought in the 12:35pm service from Yeovil Junction. It carries no smokebox shedcode but at this date the locomotive was allocated to Nine Elms. *(Leslie Freeman)*

30902 was later that same day despatched to Yeovil Junction to take over the 1:10pm stopping service from Exeter to Salisbury which Battle of Britain No 34086 *219 Squadron* had brought in. It was captured here at Templecombe with its short train where it was due to wait for some six minutes although there was no advertised connection for the S&D line at this time of day. Passengers also had to endure a five minute wait at Semley possibly to attach milk tank wagons. *(Leslie Freeman)*

The 8:10am service from Tonbridge to Brighton via Eridge rumbles over the level crossing at Barcombe Mills on 3 December 1960. The signalman standing in the door of his box seems to be well kitted out for the dismal weather apparent on that day. The route south of Uckfield to Lewes closed in 1969 although part of the route at Isfield has now been preserved as the Lavender Line. (Leslie Freeman)

On 7 August 1960 during the six month period when No 30902 was allocated to 75A it is seen parked up on Brighton shed in the company of an unidentified 3 cylinder mogul of class N1 or U1. (Transport Treasury)

30903 Charterhouse

Allocated Nine Elms 6/57 – 3/60. Guildford 3/60 – 11/62. Nine Elms 11/62. Withdrawn 12/62. Stored Nine Elms 12/62 – 2/64. Scrapped Eastleigh Works 2/64.

On 23 February 1963 some months after withdrawal No 30903 is seen at Nine Elms shed adjacent to Collett 57XX class pannier tank No 4681 which had arrived here in October 1959 taking up carriage stock duties between Waterloo and Clapham yard although it moved to Feltham shed a couple of months after this image was taken. Having lost its *Charterhouse* nameplate the Schools bears a crudely chalked replacement somewhat enigmatically styled "Less Delight" which has been applied to the backing plate. It was to linger here until early 1964 when it was towed to Eastleigh Works for scrapping. *(A. E. Bennett)*

On 27 February 1960 Redhill Shed yard witnesses No 30903 about to receive a tender full of coal from the adjacent coaling plant. Also in view amongst a number of Maunsell moguls is N Class No 31870 a resident here since the previous June together with the unmistakeable outline of a Class Q1. As the headcode discs on *Charterhouse* indicate members of the Schools class were employed at this time on cross country services from Reading through Guildford and Redhill to Tonbridge. *(R. C. Riley)*

30904 Lancing

Allocated Basingstoke 6/58. Withdrawn 7/61. Stored Ashford Works 7/61 – 8/61. Scrapped Ashford Works 9/61.

Right: On 13 May 1961 No 30904 departs Waterloo with a Southampton service towards the end of its working life being withdrawn a couple of months later in July. The proceedings are observed by a couple on the platform the gentleman deciding to take advantage of a "sit down" on the handy luggage trolley. *(A. E. Bennett)*

Left: Farnborough in October 1960 witnesses the departure of a stopping service down the ex LSWR mainline to Basingstoke hauled by *Lancing*. Of note is the grass bank topped with trimmed bushes occupying the site of the former fast line platforms which were dispensed with upon quadrupling of the line. (*A. E. Bennett*)

30905 Tonbridge

Allocated Basingstoke 6/58. Withdrawn 12/61. Stored Eastleigh Works 1/62. Scrapped Eastleigh Works 3/62.

Right Top: Labouring on a lengthy up freight near Basingstoke is a rather grimy *Tonbridge* captured here on 4 November 1961 just a month away from withdrawal. In 1958 it was allocated the unique high sided self trimming tender formerly fitted to No 30932. The donor locomotive was finished in black livery but for some reason its tender was erroneously repainted in green whilst in Ashford Works and as No 30905 was about to receive green livery it was decided to reallocate the tender. (*Charlie Verrall*)

30906 Sherborne

Allocated Nine Elms 6/57 – 3/60. Guildford 3/60 – 11/62. Brighton 11/62. Withdrawn 12/62. Stored Eastleigh Works 1/63 – 3/63. Scrapped Eastleigh Works 4/63.

Right Bottom: Also seen by the coaling stage at Redhill shed is *Sherborne* on 29 December 1960 having recently received a light intermediate overhaul at Ashford Works where, amongst other items, a speedometer had been fitted. This was a Guildford based locomotive at this date one of three, along with 30903 and 30909, that had been transferred from Nine Elms to 70C in March 1960, to take over Reading – Redhill duties, as 70A found itself with an "embarrassment of riches" with a large number of locomotives having been declared surplus by the South East division following Stage 1 of the Kent Coast electrification in the summer of 1959. The following month a further trio of Schools, Nos. 30914/15/16, moved to Redhill to assist on these workings in addition to handling the 5:25pm service from London Bridge to Reading. (*Ron Ward*)

With the ex GWR Mainline East signalbox visible in the background adjacent to the mainline from Paddington and controlling the spur down to the SR lines No 30906 rests amongst the piles of ash outside Reading South shed after having brought in a service from Guildford on 23 February 1962. *(Nick Nicolson)*

On an unrecorded date in September 1961 Clapham Junction sees the passage of No 30906 with a down service passing an up train with carriage boards proclaiming "Waterloo-Southampton-Weymouth". The SR had its own unique series of locomotive headcodes being various combinations of one, two or three discs usually with the duty number pasted onto one of the discs. *(Eric Sawford)*

30907 Dulwich

**Allocated Nine Elms 6/57 – 11/60.
Brighton 11/60. Withdrawn 9/61.
Stored Ashford Works 7/61 – 8/61.
Scrapped Ashford Works 9/61.**

Rounding the curve, which had a speed restriction of 20mph, at the west end of Fareham station and passing the signalbox comes No 30907 with the service from Bournemouth to Brighton in July 1961. A Brighton based locomotive at this date it would only have another two months of active life before withdrawal in September that year. *(J. A. C. Kirke)*

Waterloo hosts the 12.54pm service to Basingstoke powered by No 30907 adjacent to 2 HAL No 2660 on 3 February 1960. The lifespan of these two forms of motive power were to prove very similar as these EMUs ordered in 1938 began to be withdrawn some 30 years later, much like the Schools, in 1969 displaced by new 4 VEP units on main line stopping services and 4 COR units on the Reading line. The final units were to be found working on the 'Coastway' routes, as they were termed by the marketing department, from Brighton in tandem with 2 BIL units but all had been withdrawn by the end of 1971. *(Nick Nicolson)*

30908 Westminster

**Allocated Stewart's Lane until 6/59.
Basingstoke 6/59. Withdrawn 9/61.
Stored at Eastleigh Works 9/61-10/61
Scrapped Eastleigh Works 10/61.**

Basingstoke was the home depot of No 30908 and the station here is the setting for this view taken on 29 April 1961 the locomotive displaying duty No 234 indicating a local service from Waterloo. A couple with three children, two of which occupy a pram whilst the third clutches the handle tightly being no doubt unused to such a steaming beast which has just arrived from the east, observe proceedings from the up platform. *(Nick Nicolson)*

30909 St. Paul's

**Allocated Nine Elms until 3/60.
Guildford 3/60. Withdrawn 2/62.
Stored Ashford Works 2/62-3/62.
Scrapped Ashford Works 3/62.**

No 30909 is seen at Ashford on 6 November 1960 shortly before it entered the Works for what was to prove its final overhaul which took place between the 16 November and 17 December during which a speedometer was fitted. It was originally intended to undertake just a "light casual" overhaul but further work was found necessary and this was upgraded to "heavy casual" which would see the locomotive through until withdrawal in February 1962 with *St. Paul's* returning to Ashford to be cut up the following month. *(John Bell)*

Left: And here it is on the scrapline at Ashford Works yard in company with other class members which included 30900 on 25 February 1962.
In all fourteen Schools were scrapped at Ashford whilst Eastleigh Works handled twenty. *(Leslie Freeman)*

30910
Merchant Taylors

Allocated Nine Elms until 12/61.
Stored Eastleigh Works 1/62.
Scrapped Eastleigh Works 1/62.

Right: On 1 July 1960 the signalman can be seen in the isolated signalbox at Polhill situated between Knockholt and Dunton Green as he witnesses the passage of No 30910 with an up service to London Bridge from Dover operating via Ashford and Tonbridge. The train will shortly enter Polhill tunnel. The box here was to an SER design and dated from the 1870s. The serenity of this rural image is today somewhat compromised by the M25 which thunders past close to this location. *(Dave Clarke)*

A number of pedestrians, including two small boys in their school blazers and caps, walking beside the mainline glance at the sight of No 30911 heading an up service to Waterloo to the south west of Wimbledon station on 6 August 1960. The goods yard can be seen to the left and is provided with a water crane. *(Dave Clarke)*

30911 Dover

Allocated Nine Elms until 11/60. Brighton 11/60-1/62. Redhill 1/62-12/62. Brighton 12/62. Stored Hove Goods Yard 12/62-7/63. Scrapped Eastleigh Works 9/63.

Right Top: Brighton based *Dover* heads the through service from Walsall to Hastings on 12 August 1961 as the train enters the deep cutting on the approach to Patcham to the north of Brighton. This was one of no less than five through expresses serving the Sussex and Kent coastal resorts that summer, one starting from Manchester serving Sheffield, Nottingham and Leicester on Fridays only whilst another originated from Leicester on Saturdays only. A fourth through train travelled via the Great Central route from Sheffield Victoria calling at Chesterfield Central, Nottingham Victoria, Loughborough Central, Leicester Central, Rugby Central and Woodford Halse. The Kent coast resorts including Margate and Ramsgate were served by a through train from Derby Friargate operating via the GCR route through Nottingham and Leicester. No 30911 was to go on to provide one of the last sightings of a Schools on the SR's Central Division when 30911 powered the 07:27 Reading to London Bridge on 28 December 1962 then taking over the 16:40 London Bridge to Brighton via Oxted service. The following day it was unceremoniously despatched from Brighton shed to the dumping ground at Hove Goods Yard in company with 30901, 30915 and 30923. *(Brian Wadey)*

Right Bottom: With a good head of steam issuing from the safety valves No 30911 heading the Wolverhampton (Low Level) to Margate through service enters Farnborough's North Camp station on 19 April 1962 at 2pm if running to time. Although these days a bridge carries most road traffic over the railway the level crossing remains in situ but with lifting barriers. A member of the footplate crew of an unidentified Class Q1 waiting in the siding to the right walks back to his mount billycan in hand. *(Larry Fullwood)*

Minus connecting rods and nameplate No 30911 waits on the scrapline at Eastleigh on 27 July 1963 having recently arrived from several months' storage in Hove goods yard. Behind is another casualty of the cull of steam locomotives enacted at the end of the previous year in the shape of K Class mogul No 32338 another recent resident of Hove goods yard. Also in view is USA tank No 30073 still very much active at this time and indeed lasting in service until January 1967 the final year of SR steam. *(Nick Nicolson)*

30912 Downside

Allocated Nine Elms 6/59. Withdrawn 11/62. Scrapped Eastleigh Works 12/62.

The photographer's notes accompanying this image timed the shot very accurately to 8:45am on the morning of 9 April 1962. Nine Elms based Schools *Downside* has charge of a van train destined for Clapham yard and is seen here at the Junction. In 1961 this was one of the pair of the class, along with No 30921, to receive an 8 wheel tender from a withdrawn Lord Nelson class with the aim of improving braking power when handling freight services. As the Schools were not really suited to such work and as withdrawals were well under way by then anyway no further tender transfers were made. *(James Harrold)*

30913 Christ's Hospital

Allocated Nine Elms 6/59. Withdrawn 1/62. Scrapped Eastleigh Works 2/62.

After transfer from the south east Schools were regular visitors to the ex LSWR mainline from Salisbury occasionally reaching Exeter and often getting as far as Yeovil whose Junction station here plays host to No 30913 on 8 August 1960. This five coach train is standing at the former down platform at the junction which was taken out of use in 1967 when much of the route was short sightedly singled. *(Terry Gough)*

30914 Eastbourne

Allocated Brighton 6/59. Redhill 5/60. Withdrawn 7/61.

In the pleasant Surrey countryside near Betchworth on a glorious summer's day Saturday 13 August 1960 No 30914 has charge of a service from Reading to Redhill. It is obviously washing day for one of the households in the properties to the left of the line but thankfully the Schools is not making too much smoke. The slopes of the North Downs can be seen on the right and today Betchworth often plays host to steam once again as it is on the route of the popular "Surrey Hills" specials and I can recall enjoying a trip behind No 34027 *Taw Valley* along this line several years ago. *(Dave Clarke)*

The Wolverhampton to Margate and Eastbourne through train rounds the curve from the Reading line on the approaches to Guildford on 19 July 1960. *Eastbourne* was one of the class allocated to Redhill shed for working trains between Reading and Redhill but also to handle these inter-regional services which, as seen here, often loaded to 11 coaches. The correct headcode disc of one disc below the chimney, indicating a service from Reading to Margate, is carried. The junction here split into three routes, that to Reading on the left, to Woking in the middle and to Leatherhead and Epsom on the right. *(Brian Wadey)*

30915 Brighton

Allocated Brighton 6/59. Redhill 5/60. Brighton 11/62. Withdrawn 12/62. Stored Hove Goods Yard 1/63 – 6/63. Scrapped Eastleigh Works 11/63.

This is the scene at Wokingham on 28 June 1962 as a very grubby No 30915, fitted with a multiple jet blastpipe and rather ungainly wider chimney, heading the Wolverhampton to Dover and Sandwich/Eastbourne service passes through, its first stop on this leg of the journey being at North Camp. At the other platform can be seen the rear end of coach set No 451 one of the Maunsell 59' Restriction "1" 4 Corridor sets dating from 1929 which had low windows and six compartment brake vehicles. Later that year this set along with some of the others was reduced to two coaches with new 2 car sets formed from the released vehicles. No 451 lasted in service until withdrawal in May 1964. In the distance can be seen the spire of the Grade II listed St. Paul's church. The footbridge here remains in situ today for pedestrians wishing to cross the line but the original station building has long since been replaced by a modern structure. *(Peter Pescod)*

Left: The 5.25pm departure for Guildford and Reading is seen at London Bridge on 20 September 1961 with *Brighton* providing the motive power. This was one of the last of the class to be scrapped "in house" as it lingered on at Eastleigh Works until despatched in November 1963 almost a year after withdrawal. *(James Harrold)*

Right: 10 June 1962 saw a Whit Sunday excursion from Oxford to Brighton make its way past the sylvan setting of Copyhold Junction to the north of Haywards Heath with No 30915 at the head. Falling on the seventh Sunday after Easter Whitsun was a popular date for excursions by train as the weather by this date was often predictably good and the south coast resort of Brighton was perennially popular. *(Brian Wadey)*

30916 Whitgift

Allocated Brighton 6/59. Redhill 5/60. Brighton 12/62. Withdrawn 12/62. Stored Hove Goods Yard 1/63-8/63. Scrapped Eastleigh Works 9/63.

Left: This 1961 view was taken on Reading South shed where Redhill (75B) based No 30916 was being serviced having arrived on a service from Redhill and Guildford. Also on shed that day was one of Maunsell's moguls seen on the far right which shared duties with the Schools over this route. *Whitgift* lasted in service until the "Great Cull" at the end of 1962 and was another of the class to spend time stored in Hove Goods yard until despatched to Eastleigh Works in August 1963. *(Alec George)*

Right Top: Traversing the level crossing situated at the Reading end of North Camp station on 13 October 1961 No 30916 has charge of the through service from Wolverhampton to the south coast, a regular duty for these powerful 4-4-0s at this time. Originally built with four broad platforms with local army needs in mind, today the station has just two however the original buildings on the Reading platform survive albeit that only part remains in railway use whilst the main pavilion is occupied by a dentist's surgery. *(Peter Pescod)*

30917 Ardingly

Allocated Brighton 7/59. Withdrawn 11/62. Stored Brighton shed 11/62-2/63. Scrapped Eastleigh Works 4/63.

Right Bottom: On 7 August 1960 a sparkling No 30917 is seen on Brighton shed in company with an unidentified K Class mogul and one of the depot's ubiquitous E4 tank locomotives No 32475. *Ardingly* was a popular locomotive with Brighton crews and could often be seen on one of the through services from Brighton to Bournemouth or as far as Salisbury when operating the Cardiff service and occasionally the Plymouth train when one of 75A's temperamental Bulleid pacifics was unavailable which was quite often the case.
(John Robertson)

There were some equally ruinous looking locosheds on the SR by the 1960s particularly Eastbourne and some of the buildings at Bricklayer's Arms but the photographer's notes indicate that this view of *Ardingly* was taken on an unrecorded date in 1961 at Ashford. However, I still incline to the view that this is in fact the remains of Eastbourne shed which although not formally closed until June 1965 spent the last few years of its existence in the roofless condition seen here. If it proves to be Eastbourne then the Schools had probably worked in on one of the through services from the Midlands and would be serviced here. Following the closure of Brighton shed in 1964 and the rerouting of the Glasgow Car Sleeper service to Newhaven, Eastbourne shed played host to a variety of Willesden Black 5's and on one occasion 'Jubilee' 45672 *Anson*, that provided the motive power for this inter-regional service. *(Derek Potton)*

Below: The nameplate of *Ardingly* was presented to the eponymous college situated in West Sussex after withdrawal and today adorns a wall of the staff room. Several other nameplates from the class were also presented to the various educational establishments after which they were named including Sherborne, Shrewsbury, Marlborough and Haileybury. *(Derek Potton)*

30918 Hurstpierpoint

Allocated Nine Elms 6/59. Brighton 5/60. Nine Elms 11/60. Withdrawn 10/61. Stored Eastleigh Works 10/61-12/61. Scrapped Eastleigh Works 12/61.

Photographed in the unusual location of Willesden shed is No 30918 seen here being watered on 16 July 1960 in the company of a couple of diesels, one of which can be identified as Type 2 D5078 which was allocated to 1A from April 1960 until April 1966. The Schools had worked in over the West London line as indicated by the pattern of the three headcode discs whose description was given in the Ian Allan ABC as "To LMR via West London line". Schools had also been "bedfellows" with these diesels on their home turf during the previous year in Kent when the delivery of diesel locomotives, mainly for freight work on the Kent Coast and Hastings lines as part of the Kent Coast Electrification scheme, was delayed resulting in the loan of 15 Type 2s from the LMR.
(Nick Nicolson)

Adjacent to what appears to be a timber yard Basingstoke shed regularly serviced locomotives from the WR reaching the town via the route from Reading and one of their 4-6-0s is seen behind *Hurstpierpoint* in this May 1961 view. *(Derek Potton)*

30919 Harrow

Allocated Nine Elms 6/59. Brighton 5/60. Withdrawn 2/61. Stored Ashford Works 2/61-3/61. Scrapped Ashford Works 3/61.

Left Top: Images of 30919 *Harrow* during our period are few and far between perhaps because it was one of the first of its class, along with No 30932, to be withdrawn as early as February 1961 and cut up very promptly the following month. Although this image is undated the locomotive carries a Brighton shedplate and as it was allocated to 75A only from May 1960 until withdrawal it is a fair assumption that this photograph falls within our timescale. It is seen here in quite reasonable external condition with little sign of its impending demise at Brighton's impressive terminus. *(Charlie Verrall)*

30920 Rugby

Allocated Stewart's Lane 6/59. Brighton 2/61. Withdrawn 12/61. Stored Ashford Works 12/61- 1/62. Scrapped Ashford Works 1/62.

Left Bottom: Scurrying along light engine near St. Mary Cray Junction on 2 July 1960 No 30920 makes haste away from the capital at this important busy multi-tracked junction where the former SER lines to Tonbridge and Hastings and to Rochester and the Kent coast intersect and diverge with Chislehurst, Bickley, Orpington and St. Mary Cray junctions located at the four points of this complex. *(Dave Clarke)*

Right: A more tranquil setting is afforded by this view of a hikers' excursion from Victoria to Edenbridge on 21 August 1960 headed by *Rugby*. The train has almost reached the former LB&SCR station at Edenbridge Town one of two stations that served the settlement the other being located on the former SER route from Redhill to Tonbridge. Special trains for hikers remained popular well into the 1970s and gave Londoners a chance to get some fresh air away from "The Smoke". *(Dave Clarke)*

30921 Shrewsbury

Allocated Stewart's Lane 6/59. Nine Elms 8/61. Withdrawn 12/62. Stored Nine Elms 12/62-3/64. Scrapped Cohen's Kettering 4/64.

Seen running back light engine down the 1 in 53 gradient from Somerhill Tunnel to Tonbridge on 16 June 1961 Stewart's Lane allocated No 30921 displays the six wheel tender (No 710) which it has carried since August 1957. This would be replaced with a tender from Lord Nelson No 30854 in November 1961.
(Transport Treasury)

The difference is clear in this later image of *Shrewsbury* taken at Basingstoke on 25 August 1962. The tender is longer than the locomotive giving it a rather ungainly appearance but thankfully only two Schools were adorned with this replacement. *(Ron Smith)*

30922 Marlborough

Allocated Stewart's Lane 6/69. Brighton 11/61. Withdrawn 12/61.Scrapped Ashford Works 12/61.

No 30922 makes a fine sight as it emerges from Patcham tunnel with its inter-regional service returning from Hastings to Manchester Piccadilly . Whilst these through trains avoided the need to change trains, no doubt a boon to passengers with holiday luggage, they were something of a marathon with the journey from Hastings taking Mancunians home lasting 6 hours and 25 minutes! *(Brian Wadey)*

No 30922 pauses at Clapham Junction with an up service in June 1960. Of note are the typical SR hexagonal lampshades mounted on concrete posts a common sight at many stations. *(Transport Treasury)*

30923 Bradfield

Allocated Stewart's Lane 6/59. Bricklayer's Arms 5/61. Brighton 11/61. Withdrawn 12/62. Stored Hove Goods Yard 12/62-5/63. Scrapped Eastleigh Works 8/63.

At "full chat" No 30923 passes Wivelsfield on 6 August 1962 with the 10:50am Wolverhampton to Brighton and Hastings service. This train travelling via Banbury and Oxford conveyed a miniature buffet service. At Keymer Junction the train will take the Lewes line as this is probably the four coach portion destined for Eastbourne and Hastings. *(Charlie Verrall)*

The location given away by the two large water softening towers and the position of the depot clock, Brighton shed plays host to *Bradfield* seen here with coal piled high in the tender parked next to classmate No 30901 *Winchester* on an unrecorded date in 1962. No 30923 would be another of the class to be put into storage at Hove Goods yard prior to transfer to Eastleigh for scrapping. *(Barry Richardson)*

30924 Haileybury

Allocated Bricklayer's Arms 6/52. Redhill 8/61. Withdrawn 1/62. Scrapped
Ashford Works 1/62.

No 30924 is seen here on Dover shed on 2 April 1960 having worked
down from London during the time it was allocated to Bricklayer's Arms
prior to its move to Redhill. It is wearing the attractive green livery
which had been applied the previous year whilst in Ashford Works for a
general overhaul. Also in view is C Class No 31150 on the left whilst
behind the Schools is another example of these Wainwright 0-6-0s,
Dover having six of them on its books at this time. *(David Idle)*

Taken shortly before the image forming the frontispiece to this volume this angle shows the line of imported cars together with the French Railways van heading the consist hauled by *Haileybury*. The cargo seems to be attracting the attention of a couple on the up platform whilst standing at platform 3 is Hastings 6B unit No 1032 constructed in 1958 and, as the designation suggests, originally containing a buffet car which to effect economies was withdrawn from this set in 1964 being replaced by a 6S trailer. *(Dave Clarke)*

No 30924 is seen at Ashford shed on 24 April 1960. A wide variety of motive power is seen in the right background including an H Class tank and a Q1. There are also a couple of diesel shunters in view, the depot having examples of classes 03, 08 and 12 based there. Ashford also had half a dozen Schools of its own at this time but *Haileybury* was not one of them. *(Nick Nicolson)*

Left: This is believed to be the last steam hauled 9.10am service from Charing Cross to Folkestone, Dover and Ramsgate seen here entering Waterloo's Eastern section platforms on 10 June 1961 with No 30924 doing the honours. *(Leslie Freeman)*

Right Top: The popular commuter town of Reigate witnesses the passage of a through service from the Midlands on 22 July 1961 with the usual motive power at this time in the shape of Schools class *Haileybury*. Advertisements on the platform include one for the latest release showing at the London Pavilion which was the low budget rather racy "A cold wind in August" starring Lola Albright as a mentally unbalanced stripper in her 30s who becomes involved in a torrid romance with a 17-year-old boy. Not sure how much this would have appealed to "Disgusted of Reigate rather than Tunbridge Wells"! The cinema went on to host world premieres of many famous movies such as "Dr. No", "A Hard Day's Night" and "Yellow Submarine". It closed in 1981 and was sold only to be later internally demolished. *(Brian Wadey)*

30925 Cheltenham

Allocated Bricklayer's Arms 5/51. Stewart's Lane 2/61. Basingstoke 8/61. Withdrawn 12/62. Preserved part of National Collection York. Currently on Mid Hants Rly.

Right Bottom: Milk from the West Country despatched from such stations as Torrington, Lapford, Seaton Junction, Chard Junction and Semley arrived in London daily and was usually destined for the large United Dairies depot at Vauxhall. Here No 30925 is seen with a train of milk tankers at Clapham Junction on 14 October 1961. *(Brian Wadey)*

With the demise of the class imminent Schools were in demand for railtours and *Cheltenham* was a regular performer on such duties. The RCTS "Sussex Special" railtour has just arrived at Brighton having brought a trainload of enthusiasts from London Bridge on 7 October 1962 with an express timing of 60 minutes for the run, although it actually took 64 minutes with a load of 7 coaches but 7 minutes were put down to delays. The tour then proceeded to Seaford and return with AIX No 32636 and E6 No 32418 double heading. K Class mogul No 32353 returned the tour to London via Preston Park, Steyning and Epsom. The route via Preston Park was necessary as there was no access to the coast line to Shoreham-by-Sea from the platform at Brighton at which the special had arrived. *(Alec Swain)*

A rare bird indeed at Nottingham Victoria as No 30925 in tandem with fellow 4-4-0 2P No 40646, whose boiler ticket had apparently expired, prepares to take out the RCTS "East Midlander No 5" railtour of 13 May 1962. Departure from Nottingham was at 8:35am and arrival back after visiting Darlington Works was not until 9:49pm with the route traversed taking in Chesterfield, Wakefield, Harrogate, Ripon and York. *(Alec Swain)*

30926 Repton

Allocated Bricklayer's Arms 5/51. Stewart's Lane 2/61. Bricklayer's Arms 5/61. Stewart's Lane 11/61. Basingstoke 1/62. Withdrawn 12/62. Preserved North York Moors Rly.

Left: This June 1961 image reveals *Repton* carrying a 73A shedcode parked up at Stewart's Lane depot adjacent to the South London line viaduct linking Victoria with Wandsworth Road. Beyond this viaduct the large coaling tower, dating from the SR rebuilding of the depot in 1933/34, can be seen. Commensurate with construction of this facility the engine shed received a replacement asbestos roof, new watering facilities were provided and the turntable moved several hundred yards to the west. Stewart's Lane was then capable of handling up to 170 locomotives which gave it the largest allocation on the SR network. Of interest is the rather primitive fuelling facility seen on the right. The line out of shot crossed by the girder span extension of the viaduct glimpsed to the extreme left is the West London Extension line an important cross capital route from Willesden Junction. *(J. A. C. Kirke)*

Below: Schools were often used on the Royal Train taking members of the family to Epsom for The Derby and No 30926 was given the honour in 1961 and is seen here on 11 June with three Pullman cars, the leading one of which is "Isle of Thanet", and a royal saloon with the locomotive sporting the appropriate headcode for the occasion. It was photographed at an unrecorded location on the route between Victoria and Tattenham Corner. For this prestigious duty a standby Schools was also readied, should it be needed. *Repton* again worked a similar train the following year whilst No 30901 acted as standby locomotive, this year of course being the final one at which a Schools officiated due to their withdrawal from traffic at the year end. Pullman car No 247 *Isle of Thanet* constructed in 1924 and originally named *Leona* was frequently to be found in Royal Trains until its last recorded appearance in 1967. Following presentation to the USA Ambassador it was shipped to the National Railroad Museum at Green Bay, Wisconsin in 1969 and accompanied Flying Scotsman on its USA tour. Repatriated to the UK in 2000 it can now be found on the K&ESR. *(Stan Creer)*

Repton was on railtour duty on 25 February 1962 when it was captured here about to leave Appledore with the LCGB "Kentish Venturer" special, a tour run to mark the end of steam in Kent. Participants had been brought from Victoria to Ashford by Class N15 No 30782 *Sir Brian* thence to Appledore courtesy of Maunsell mogul No 31690 double heading with Class H No 31263 for a trip over the New Romney branch. No 30926 would return the tour to Charing Cross having earlier in the day brought the empty stock into Victoria and performed banking duties on the steep climb up to Grosvenor Bridge. This was the second of five railtours in which Schools took part during their last year in service, the last being the Home Counties Railway Club "Eastleigh Special" of 11 November. *(Leslie Freeman)*

30927 Clifton

Allocated Bricklayer's Arms 5/51. Feltham 2/61. Nine Elms 3/61. Withdrawn 1/62. Stored Eastleigh Works 1/62-4/62. Scrapped Eastleigh Works 4/62.

An up train from the Kent coast to Charing Cross headed by No 30927 speeds through Paddock Wood on 10 September 1960. The famous signalbox straddling a siding which fed off the up platform loop afforded the signalman an unhindered view over the station footbridge and was one of a pair to an SER design which controlled the layout here. The box closed in April 1962 following the commissioning of the new Tonbridge panel box. A bay platform just off the image to the right was used by branch trains to Hawkhurst. The tall down signals on the left gave approaching drivers a good view ahead which would otherwise be impeded by the station footbridge. *(Leslie Freeman)*

Clifton approaches Wimbledon on 29 July 1961 with one of the summer Saturday boat trains to Lymington where connection could be made for the Isle of Wight ferry. In the summer of 1961 three services were timetabled to leave Waterloo at 8:45am running until 2 July, 9:42am, and 12:00 noon running from 15 July until 2 September for Lymington Pier, a change of locomotive being necessary at Brockenhurst for travel down the Lymington branch. In the reverse direction there were no fewer than five departures for Waterloo from Lymington Pier with one running only from 15 July until 2 September and one from 1 July to 2 September. These through trains had been introduced as far back as 1926 and often loaded to nine or ten bogies with restaurant car and by 1962 the Schools had come to monopolise these workings. Wimbledon "C" signalbox seen at the divergence of the route to Sutton was an SR type 11C box dating from 1929 and like many boxes in the London area was provided with a brick enclosure around the base to afford bomb blast protection during WW2. *(A. E. Bennett)*

30928 Stowe

Allocated Bricklayer's Arms 3/51.
Brighton 11/61. Withdrawn 11/62.
Preserved Bluebell Rly.

Backing onto its train at Charing Cross on 23 July 1960 is No 30928 carrying a 73B Bricklayer's Arms shedplate with a service for Dover via Tonbridge. A family with their suitcases probably packed for a seaside holiday, wait patiently for their train on the platform. Down below at street level on the left is Northumberland Avenue with the former Metropole Hotel, now the Corinthia Hotel, dominating the street scene – no vehicles in sight, very different from the capital's traffic choked roads of today. (Nick Nicolson)

The through service from Brighton arrives at its destination of Bournemouth West on 27 April 1962 with Brighton based No 30928 heading the train. (Frank Goudie)

Left: The fireman looks out from the cab of *Stowe* as it gingerly handles a goods service being subject to the 20mph speed restriction over this junction to the west of Tonbridge on 1 July 1961. *(Leslie Freeman)*

30929 Malvern

Allocated Bricklayer's Arms 8/50. Stewart's Lane 11/61. Brighton 1/62. Withdrawn 12/62. Scrapped Eastleigh Works 3/63.

Right: This nicely lit side on portrait of No 30929 at Brighton showing its clean lines and attractive profile dates from 29 October 1962 just a few days before withdrawal. In addition to its own allocation of three Schools at this time, Nos 30901, 30928 and 30929, Brighton would gather in seven further examples in the shape of 30906, 30911, 30915, 30916, 30917, 30923 and 30930 during the following two months prior to the wholesale withdrawal of the class at the end of that year. *(R. C. Riley)*

In happier times *Malvern* is seen on its old stomping ground of Kent as it heads a down service from London Bridge to Dover into Dunton Green on 18 April 1960. As the large running in board indicates, passengers could change here for Chevening, Brasted and Westerham services which departed from the platform on the far left. The signalbox was a standard 50 lever SE&CR box based on a Saxby & Farmer design but due to its position on the far side of the main line the signalman was required to cross the main lines repeatedly with the single line tablet for the Westerham branch. To avoid this hazardous manoeuvre a system of cables and pulleys was devised which carried the tablet from the signal box to an elevated stage which could be accessed by branch line enginemen via some wooden steps at the north end of the branch line platform. *(Nick Nicolson)*

On 31 May 1962 No 30929 was reduced to acting as a lowly station pilot at Redhill where it is seen in between duties. *(Leslie Freeman)*

30930 Radley

Allocated Bricklayer's Arms 6/57. Stewart's Lane 11/61. Brighton 1/62. Redhill 2/62. Brighton 12/62. Withdrawn 12/62. Stored at Redhill 1/63-3/64. Scrapped Eastleigh Works 4/64.

Two views of No 30930 at Reading General taken on Saturday 21 July 1962. In the first view the locomotive gets some admiring glances from a gaggle of platform "spotters" whilst waiting to take over an inter regional service from the Midlands. In the second view *Radley*, carrying the single disc headcode for Reading to Margate via Redhill, is impatient to be off having coupled up to its train whilst a Castle class locomotive is seen at the opposite platform carrying reporting No 040 which is believed to be a Weston-super-Mare to Paddington service. On the far left an Ian Allan ABC is being consulted by one of the locospotting fraternity. (*Nick Nicolson*)

This view of Bricklayer's Arms based No 30930 seen at the coaling tower at Ramsgate depot in company with Standard tank No 84021 dates from 25 August 1960. The class 2 tank had been delivered new to Ashford depot 3 years previously and continued to be based there until transfer to Bricklayer's Arms in February 1961. *(Terry Gough)*

Seen towards the end of its working life at Redhill shed, the date being 19 September 1962, No 30930 would be formally withdrawn at the end of December although it looks as if it has already been taken out of traffic at the time of this image. It would go on to spend the next 15 months stored at 75B until carted off to Eastleigh Works for scrapping in April 1964. *(Ron Ward)*

30931 King's Wimbledon

Allocated Bricklayer's Arms 4/46. Feltham 2/61. Nine Elms 3/61. Withdrawn 9/61. Scrapped Ashford Works 10/61.

The unmistakeable facade of Feltham shed provides the backdrop to this view of No 30931 seen here together with classmate No 30939 on 5 March 1961. These two Schools together with No 30927 were only allocated very briefly to 70B in the early part of 1961 being shunted off to Nine Elms, who already had more than enough on their books, when Feltham decided that they could not usefully make use of them. *(Brian Wadey)*

Perhaps of more interest than the freight service, which also includes SNCF ferry vans, headed by *King's Wimbledon* is the wreck of Bulleid Pacific No 34084 *253 Squadron* seen here on its side after coming to grief following a shunting accident. Whilst hauling an early morning Dover Marine to Bricklayer's Arms van train freight it overran signals and became derailed toppling down an embankment. The tender was recovered on 24 February with the locomotive following on 28 February. The strange sight of the stranded Pacific gets more than a passing glance from a member of the footplate crew as his steed passes Hither Green on 21 February 1960. *(R. C. Riley)*

30932 Blundell's

Allocated Ashford 6/59. Withdrawn 2/61. Stored Ashford Works 6-7/61. Scrapped Ashford Works 8/61.

Left Top: Bereft of nameplates a very forlorn looking 30932 *Blundell's* rests on the scrapline at Ashford awaiting its fate on 16 April 1961. With No 30919 it was one of the first pair of its class to be withdrawn, from Ashford shed the month before, it would be scrapped in the works here in August one of four of the class still wearing black livery upon withdrawal. The patch fitted to its right hand cylinder cover during its general overhaul in 1955 can still be seen. *(Nick Nicolson)*

30933 King's Canterbury

Allocated Ashford 6/59. Nine Elms 10/61. Withdrawn 12/61. Scrapped Ashford Works 12/61.

Left Bottom: On 30 July 1960 an inter regional service from the Midlands headed by No 30933 takes the down through road at Tonbridge although much to the obvious frustration of the driver it has been brought to a stand by an adverse signal. *(Dave Clarke)*

Right: Overhauling an EMU, headed by 2 EPB No 5738 operating route 16 from Charing Cross to Sevenoaks via Orpington, and photographed near Orpington with a down service from Charing Cross on 31 May 1960 comes *King's Canterbury*. This locomotive was rumoured to be the subject of a preservation proposal by Canterbury Corporation but whether this was just speculation by railway enthusiasts or not the rumour did appear in the railway press at the time but it proved to be unfounded and the locomotive went for scrap at the end of 1961. *(Dave Clarke)*

Left Top: A similar down service from Charing Cross headed by the same locomotive No 30933 passes the EMU carriage sheds at Orpington a month later on 30 June 1960. This impressive structure was 1050 feet long becoming the largest carriage shed on the SR and consisted of a pitched roof steel frame clad with asbestos. Each of the 4 roads was capable of accommodating thirty two vehicles. The unmistakeable rear end of a Class Q1 tender can be seen manoeuvring on the left. *(Dave Clarke)*

30934 St. Lawrence

Allocated Ashford 6/59. Basingstoke 10/61. Withdrawn 12/62. Stored Basingstoke 12/62-3/63. In steam to Eastleigh 3/63 then in store 4/63 -7/63. Scrapped Eastleigh Works 8/63.

Left Bottom: The neat nameplates of the Schools class are epitomised in this close up view of No 30934 one of three of the class that featured saint or rather the shortened "ST." in their names. The rendering of ST. did vary however as Nos. 30909 and 30934 had a half height upper case T set low followed by a full stop whereas No 30938 was unique in having a half height T set high rather than low and not followed by a full stop. Minutiae beloved of "rivet counters" perhaps but an intriguing snippet of information nonetheless. *(Terry Gough)*

Right: *St. Lawrence* hurtles through Paddock Wood on 10 September 1960 with an up boat train to Victoria. A rake of mineral vans, a couple of which contain cable drums, is apparent on the left beyond the fine signal gantry with their lofty repeaters. As mentioned in the Introduction this was to prove to be the last Schools to turn a wheel in service when it travelled from Basingstoke to Eastleigh as late as May 1963 – a full five months after the class was supposedly banned from operations.
(Leslie Freeman)

Moving at a somewhat slower pace No 30934 is seen at Salisbury with a down freight consisting of empty milk tank wagons on 24 August 1962. Superseding the previous practice of carrying milk by rail in churns, bulk milk transportation by rail began in 1931 with the introduction of the 4 wheel milk tanker. However, these early wagons experienced issues with poor running and to alleviate these problems a 6 wheel tanker, as seen in this image, was developed for United and Express Dairies. These proved to be very successful with over 630 being constructed between 1932 and 1948 and they were to be seen all over the BR network until road tanker transport became predominant. Several examples survive in preservation including SR 4409 which can be seen at Didcot. *(Ron Smith)*

On 21 August 1960 No 30934 handling the 6:27pm relief boat train from Newhaven Marine to Victoria is seen here between Newhaven Harbour and Newhaven Town stations. "The Engineer" pub seen on the left is still in business today but the sidings seen on the right are no longer in situ. The withdrawal of services from the Marine station in 2006 on safety grounds was followed by formal closure of the station in 2020. Ferries plied between Newhaven Marine and Dieppe where connections were available for Paris. Fares from London to Paris via Newhaven in 1960 were £8 9s single 2nd. Class and £12 1s 1st. Class. In 2024 there were three sailings daily to Dieppe operated by DFDS. *(Horace Gamble)*

30935 Sevenoaks

Allocated Ashford 6/59. Nine Elms 11/61. Withdrawn 12/62. Stored Nine Elms 12/62-3/64. Scrapped Cohen's Kettering 4/64.

Left: Seen standing by Kensington North Main signalbox at Kensington Olympia on 6 August 1960 is No 30935. Like the box at the south end of the station North box had a WR lever frame but was of LNWR heritage, the WR frame dating only from 1956 apparently coming second hand from South Wales. During the 1960s the West London line came under the control of the WR thus there was a mixture of different signalling in existence at Olympia with lower quadrant signals in use at the south end and LMR upper quadrant signals in use at the north end. The terrace of houses in Russell Gardens on the left with their diversity of chimney pots had a good view of the variety of motive power that could be witnessed at this important cross London location but this aspect was probably not a selling point for the residents. However, today it has become a fashionable mews location. Could the inhabitants of this street in the 1960s ever have dreamt that in 2024 a three bedroom terraced house here would set you back £1.8m? (R. C. Riley)

Right Top: The 12:45pm service from Waterloo to Basingstoke is going well behind No 30935 on 7 June 1962 between Farnborough and Fleet. The first stop for this train after leaving Waterloo was Woking and it was due into Basingstoke at 2:14pm after having made further stops at Fleet, Winchfield and Hook. (Larry Fullwood)

Right Bottom: On 13 May 1961 Ashford based *Sevenoaks* pulls into Tonbridge with a through service from the Midlands to Margate. (Brian Wadey)

30936 Cranleigh

Allocated Ashford 6/59. Nine Elms 11/61. Withdrawn 12/62. Stored Nine Elms 12/62 – 7/63. Scrapped Eastleigh Works 11/63.

Vauxhall witnesses the passage of No 30936 which has charge of the 11:43 "Lymington Boat" service from Lymington Pier to Waterloo on Saturday 1 September 1962. These workings were something of a swansong for these 4-4-0s with 30902, 30935 and 30936 being noted as late as August of that year operating the three timetabled departures from Waterloo. Crompton diesels were also putting in appearances on these trains and were to largely take over from the now withdrawn Schools the following year. *(Leslie Freeman)*

On 27 July 1963 time was nearly up for *Cranleigh* as it (bereft of nameplates and in company with other withdrawn stock which includes Class O2 No 30199, unidentified Class B4 and USA tanks together with a three cylinder Maunsell mogul) waits for the final call at Eastleigh. *(Ron Smith)*

30937 Epsom

Allocated Ashford 6/59. Bricklayer's Arms 5/61. Nine Elms 11/61. Withdrawn 12/62. Stored Eastleigh Works 2/63-5/63. Scrapped Eastleigh Works 6/63.

Shunting vans at Clapham Junction on 11 April 1962 is locally based Nine Elms Schools 30937 *Epsom* still wearing the BR green livery with which it was finished in 1957. Like many of the class it had gone through a number of livery changes throughout its life starting with Maunsell or sage green when new, malachite green applied in 1939, black in 1945, malachite in 1948, black in 1950 and finally BR green in 1957. Just four of the class remained in black livery upon withdrawal – 30900, 30914, 30919 and 30932. *(James Harrold)*

Seen alongside one of Guildford depot's Maunsell moguls, in this case U Class No 31635, No 30937 still sporting its Ashford 74A shedplate comes off shed at 70C on 28 September 1962 preparatory to working a train to Reading. No 31635 would last in service until December 1963 but the final quartet of Guildford's moguls would continue until withdrawal in June 1966. *(Leslie Freeman)*

30938 St. Olave's

Allocated Dover 6/59. Ashford 7/60. Nine Elms 5/61. Withdrawn 7/61. Stored Ashford Works 7/61-9/61. Scrapped Ashford Works 9/61.

Seen from the vantage point of Hilders Lane overbridge No 30938 hurries along on 14 April 1960 near Edenbridge with a service to Redhill on the former SER route which here crosses over the former LB&SCR route from Eridge to Oxted. Bridge No 57 marks the crossing with the Eridge route having just emerged from one tunnel only to enter another tunnel shortly afterwards. The LB&SCR line was required to negotiate a hill in the vicinity via a tunnel and had intended to pass underneath the SER's Redhill to Tonbridge line. However, as the LB&SCR line was unable to attain a sufficient depth to bore underneath the SER the resulting tunnel was perforce built in two sections. The cutting seen here conveyed the former SER line between the two inner portals of the tunnel with the northern tunnel being 116 yards long whilst the southern was 179 yards long. *(Terry Gough)*

With the shunting disc signal "off" No 30938 backs out of Tonbridge on 3 April 1961 heading either towards the west yard or the Redhill line which diverges to the left by the former West signalbox later denoted "A" box. The new power box which opened on 18 March 1962 superseding both "A" and "B" signalboxes can be seen taking shape in the background. One of the properties on the right belongs to T. Ives & Son Ltd. who describe themselves on their advertising hoarding as cricket ball manufacturers, the local area being apparently well known for this rather obscure trade. In the early 1960s five of the local cricket ball makers merged to become Tonbridge Sports Industries later becoming known as British Cricket Balls! *(Dave Clarke)*

St. Olave's is seen at Ashford on 11 March 1961 just four months before withdrawal. Whilst many of the names of the class were well known to the general public No 30938 carried the name of perhaps one of the more obscure educational establishments which could be immediately linked to a particular geographical locality. In fact St. Olave's Grammar can be found in Orpington in Kent and now advertises itself as a highly selective boys secondary school founded by royal charter in 1571. *(Nick Nicolson)*

30939 Leatherhead

Allocated Dover 6/59. Ashford 7/60. Feltham 2/61. Nine Elms 3/61. Withdrawn 6/61. Stored Ashford Works 7/61-8/61. Scrapped Ashford Works 9/61.

Left: On 25 August 1960 No 30939 passes the four road carriage sheds after departure from Ramsgate with a London Bridge service. Heralding the new order on the Kent coast following the recent electrification scheme a number of EMUs of HAP and CEP varieties can be seen to the right and left of this view and some also occupy the carriage shed. Also in view, above the rear of the departing train is the lofty main station booking hall. This building is thought to have been designed by Southern Railway architects R. J. Scott and E. M. Fry, built between 1924 and 1926 and has been Grade II listed since 11 July 1988 due to its significant architectural interest. Margate station has a similar structure, but a third at Dumpton Park was demolished. On the far left is the six road dead end shed which closed to steam in 1959 after which it was converted for EMU use. (Terry Gough)

Right Top: Sandwiched between Class E6 No 32408 and S15 No 30833 is the last of the class numerically No 30939 which like the other two of the class briefly allocated early in 1961 is seen at Feltham in February during its short stay. The Billinton 0-6-2T would depart Feltham in May 1962 for Eastleigh where it would be scrapped as part of the cull at the end of the year. The S15, one of no less than 23 of the class shedded here at this time, would remain at 70B until withdrawal in May 1965. (Don Matthews)

Right Bottom: The junction to the east of Tonbridge with the locomotive shed visible in the right background was photographed on 27 August 1960 as *Leatherhead* approaches from the Kent coast with a London service. Another elevated signalbox, so typical of this route, controls the junction to the right which led to Tunbridge Wells and Hastings. The goods sidings on the left are well filled at this time and presided over by the large goods shed whilst coal, bagged up from the adjacent trucks, waits to be loaded onto lorries parked near the signalbox. (Dave Clarke)

The End of the Line

Truly the end of the line as this graphic image of the remains of No 30935 amongst the debris at Cohen's yard at Cransley near Kettering on 5 July 1964 testifies. The arrival of these locomotives from the far off SR was greeted with astonishment by local Midland enthusiasts who kept a watchful eye on arrivals at this Northamptonshire scrapyard. *(Eric Sawford)*